Editor's Comment

Of the 141 operable refineries in the United States- there are 30 of them with capacity over 100,000 barrels per Atmospheric Crude Oil Distillation- this edition reveals the operators of these refineries. Also of interest our report shows that for the first time in many year, North American refiners were less profitable than European refiners in second quarter 2016.

The seven region in the U. S. that accounts for 92 percent of domestic oil and gas production- what is their current production for the month of August and what is their expected production for the month of September, find out? In this edition- CO2

emissions from natural gas surpass coal as fuel use for the first time in 44 years- Why is this so?

Our Event- The international pipeline, oil and gas safety conference March 14-16, 2017, POGS '17- Registration now Open and Call for Abstract still on going. Register Today @ http://oilandgassafetyconference.com

Recognition- Our Magazine is now one of the Best Sellers on Amazon in Petrochemical, Petroleum and Oil and Energy Categories. Get a copy and read about the latest oil and gas industry news.

- Gloria Towolawi

Contents

USA Oil and Gas Monitor
A RGT Media Communications Corp.

Editor-in-Chief
Gloria Towolawi

Europe Bureau
Esther Coker

Nigeria Bureau
David Arhavbarien

Contributing Editor
Gloria Instead

Reporter
Caleb Motinwo

Advert & Marketing
Jewel Spring
T: 832-486-0095
E: advertise@usaoilandgasmonitor.com

Distribution & sales
Richard Godfirst

Subscribers Service
E: subscribe@usaoilandgasmonitor.com

RGT Media Communications Corp.
Publishers of
USA Oil and Gas Monitor
Workplace Weekly News
GlobalPRPlus

USA Oil and Gas Monitor is published 12 times a year monthly by RGT Media Communications Corp. 10777 Westheimer #1100

Houston, Texas 77042
Subscription price is $144 per year.
Digital copy $9.99 per download.

P O G S March 2017

The Intl Pipeline Oil and GAS Safety Conference and Exhibition

March 14-16, 2017 Houston Texas USA

Organized by:

RGT MEDIA COMMUNICATIONS CORP.

Pipeline Integrity | **Emission Reduction** | **Well Control** | **Oil and Gas Transportation** | **Chemical Extraction**

Connecting Supplier with Procurement Teams

Exhibition
200+
Exhibitors Expected

Attendance
2000+
Attendees Expected

Goal

Improve safety in the entire value chain of the oil and gas industry not limited to the well heads but distribution chains, transportation and supply chain.

Exhibit@ P O G S Safety Tech

P O G S Safety Tech provides international and local energy companies who operate across the up, mid and downstream sectors of the oil &gas supply chain with a B2B platform to meet and influence highly-focused International decision-makers and buyers.

Who is Attending?

Take Advantage of Early Registration- Register Now @

http://www.oilandgassafetyconference.com registration/online-registration/

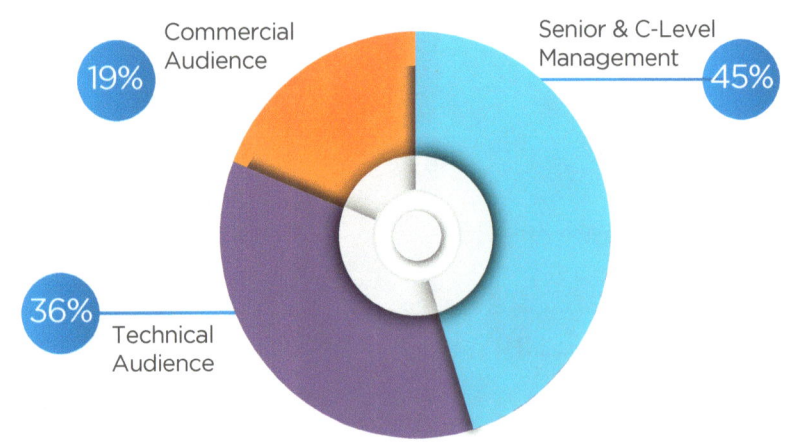

- 45% Senior & C-Level Management
- 19% Commercial Audience
- 36% Technical Audience

Who is Exhibiting?

SHOWFLOOR IS Selling Very Fast RESERVED Today

http://www.oilandgassafetyconference.com/booth-registeration/

Official Media Partner

For further details visit website @
http://oilandgassafetyconference.com
or call +1-832-664-0618

Danos' Amelia Location Authorized as Port Facility

The U.S. Coast Guard has certified Danos' fabrication facility in Amelia, La. as an official port facility. This designation allows both foreign and US vessels to access the facility or remain docked there.

The Coast Guard granted Danos port facility status following approval of the company's facility security plan under Title 33 Code of Federal Regulations CFR Part 105.

"In addition to opening our Amelia facility up to foreign vessels for transporting fabrication projects, port facility status also allows us to provide docking services for idle vessels," said Mark Danos, vice president of project services.

Located on 175 acres along Bayou Boeuf, the facility's five thousand linear feet of bulkhead and 18-foot water depth can accommodate large-scale custom fabrication projects of more than 1,000 tons.

September 2016 • Issue 9

plastics. Results of the research were published in the peer-reviewed journal Science.

If brought to industrial scale, this breakthrough could reduce industry's global annual carbon dioxide emissions by up to 45 million tons, which is equivalent to the annual energy-related carbon dioxide emissions of about five million U.S. homes. It could also reduce global energy costs used to make plastics by up to $2 billion a year.

Using a molecular-level filter, the new process employs a form of reverse osmosis to separate para-xylene, a chemical building block for polyester and plastics, from complex hydrocarbon mixtures. The current commercial-scale process used around the world relies on energy and heat to separate those molecules.

"Through collaboration with strong academic institutions like Georgia Tech, we are constantly exploring new, more efficient ways to produce the energy, chemicals, and other products consumers around the world rely on every day," said Vijay Swarup, vice president of research and development at ExxonMobil Research and Engineering Company. "If advanced to commercial-scale application, this technology could significantly reduce the amount of greenhouse gas emissions associated with chemical manufacturing."

ExxonMobil and Georgia Tech Innovation Could Lead to Significant Cuts in Chemical Manufacturing Energy Use and Emissions

ExxonMobil

- Molecular-level filter could revolutionize energy-intense chemical process
- Significantly reduces amount of energy used in polyester and plastic manufacturing
- Research published in nation's leading peer-reviewed journal, Science

Scientists from ExxonMobil and the Georgia Institute of Technology have developed a potentially revolutionary new technology that could significantly reduce the amount of energy and emissions associated with manufacturing

The research successfully demonstrated that para-xylene can be separated from like chemical compounds known as aromatics by pressing them through a membrane that acts as a high-tech sieve, similar to a filter with microscopic holes. Commercially practiced separations involve energy-intensive crystallization or adsorption with distillation. Globally, the amount of energy used in conventional separation processes for aromatics is equal to about 20 average-sized power plants.

The ExxonMobil and Georgia Tech team first developed a new carbon-based membrane that can separate molecules as small as a nanometer. The membrane was then incorporated into a new organic solvent reverse osmosis process, during which aromatics were pressed through the membrane, separating out para-xylene.

"In effect, we'd be using a filter with microscopic holes to do what an enormous amount of heat and energy currently do in a chemical process similar to that found in oil refining," said Mike Kerby, corporate strategic research manager at ExxonMobil.

The carbon-based membrane developed by the ExxonMobil-Georgia Tech team is about 50 times more energy efficient than the current state-of-the-art membrane separation technology. Because the new membrane is made from a commercially available polymer, ExxonMobil believes it has potential for commercialization and integration into industrial chemical separation processes.

Reverse-osmosis membranes are already widely used to desalinate seawater, consuming a fraction of the energy required by thermally driven processes. The new organic solvent reverse osmosis process is believed to be the first use of reverse osmosis with carbon membranes to separate liquid hydrocarbons.

"By applying pressure at room temperature, the membrane is able to concentrate para-xylene from a mixture at high rates and low energy consumption relative to state-of-the-art membranes," said Ryan Lively, an assistant professor in Georgia Tech's School of Chemical & Biomolecular Engineering and the lead researcher. "This mixture could then be fed into a conventional thermal process for finishing, which would dramatically reduce total energy input."

The technology still faces challenges before it can be considered for commercialization and use at an industrial scale. The membranes used in the process will need to be tested under more challenging conditions, as industrial mixtures normally contain multiple organic compounds and may include materials that can foul membrane systems. The researchers must also learn to make the material consistently and

demonstrate that it can withstand long-term industrial use.

"The implications could be enormous in terms of the amount of energy that could be saved and the emissions reduced in chemical and product manufacturing," said Benjamin McCool, an advanced research associate at ExxonMobil and co-author of the research. "Our next steps are to further the fundamental understanding in the lab to help develop a plan for pilot plant-scale demonstration and, if successful, proceed to larger scale. We continue to work the fundamental science underlying this technology for broader applications in hydrocarbon separations."

Chemical plants account for about eight percent of global energy demand and about 15 percent of the projected growth in demand to 2040. As global populations and living standards continue to rise, demand for auto parts, housing materials, electronics and other products made from plastics and other petrochemicals will continue to grow. Improving industrial efficiency is part of ExxonMobil's mission to meet the world's growing need for energy while minimizing environmental impacts.

The researchers on the technology as written in Science include Lively and Dong-Yeun Koh from Georgia Institute of Technology and McCool and Harry Deckman from ExxonMobil.

September 2016 • Issue 9

API- New Report Confirms Pipelines Are One of the Safest Ways to Transport Energy

API has released a new report, with the Association of Oil Pipe Lines AOPL, confirming that pipelines continue to be one of the safest ways to efficiently transport energy across the United States. The report highlights liquid pipeline safety performance and industry-wide efforts to improve pipeline safety in 2016 and beyond.

"Safety is a core value throughout our industry," said API Pipeline Manager David Murk. "This report shows the tremendous priority we place on pipeline safety, but as an industry we can always do more. By constantly evaluating our safety programs and activities, learning from past experiences, and making timely and adequate adjustments, our industry will continue working towards its goal of zero incidents." Despite a 13 percent increase over the last five years in miles of pipeline delivering crude oil, petroleum products and natural gas liquids, pipeline incidents per mile larger than 500 barrels are down nearly a third. Incidents potentially impacting people or the environment outside of an operator's facility are down 52 percent since 1999.

The report summarizes industry-wide proactive pipeline safety principles; provides a transparent analysis of industry safety records, including where performance is improving and where challenges remain; and outlines the significant efforts liquid pipeline operators continue to make in advancing technologies and implementing innovative approaches to inspecting, monitoring, and managing pipeline safety programs.

"Pipelines are a vital part of this nation's energy infrastructure and ensuring they continue to operate safely will be critical to securing our energy future," said Murk. "Plus, the construction of new pipelines will create jobs, grow the economy and guarantee all Americans are benefitting from our nation's energy renaissance."

Experience the Future of Design with ShipSpace™

The ShipSpace™ product is a next-generation Virtual Reality VR Design Verification Tool that allows designers, engineers and key stakeholders to validate design ideas and communicate effectively about vessel concepts and designs.

The product consists of computer hardware and software on the customer's premises and in the cloud.

KNUD E HANSEN is showcasing ShipSpace™, a new Virtual Reality VR tool to aid ship designers and engineers, at the SMM in Hamburg. Interested parties will be able to enter ShipSpace™ themselves and try the cutting edge design verification tool to better understand the ship design and the spaces within it.

ShipSpace™ uses the latest VR technologies to enable users to walk around all areas of the vessel and get a better understanding of how spaces work, with a true sense of depth and scale, not possible with monitors or projectors.

Finn Wollesen Petersen, Managing Director said, "ShipSpace™ is a cutting edge tool that our engineers and partners can use to design better ships, faster. KNUD E HANSEN has been an

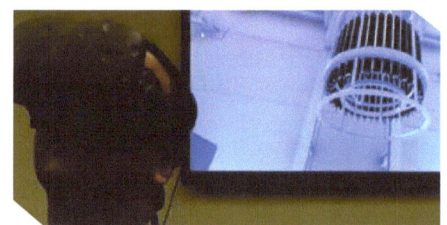

innovator in ship design since 1937 and ShipSpace™ is another step forward for us."

"ShipSpace™ has proven to be an excellent tool for communicating and developing designs and ideas, by enabling stakeholders to walk around the vessel while it is still on the drawing board" added Robert J Spencer, Head of Simulation Products.

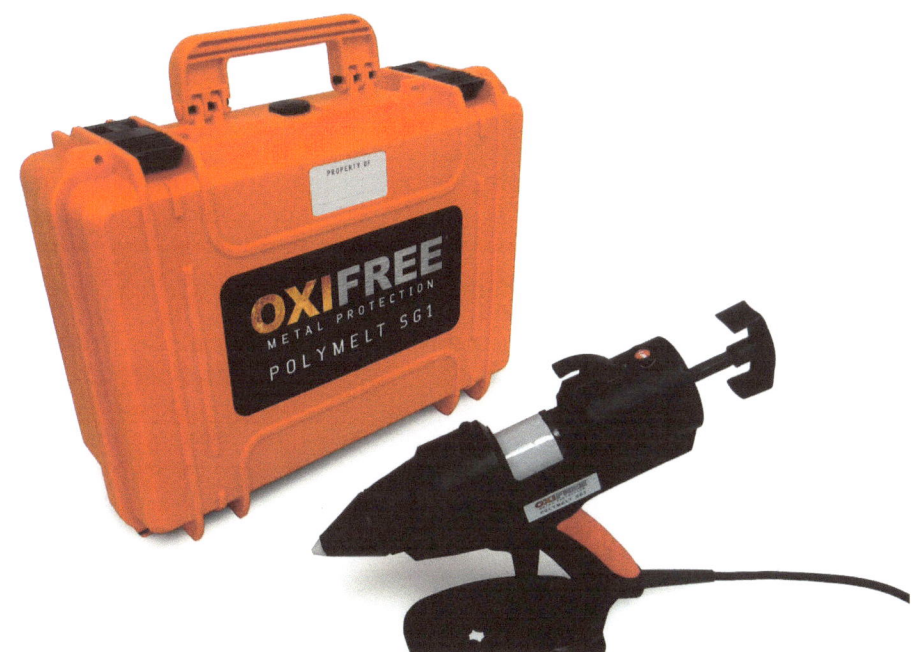

Oxifree Global unveils Polymelt Service Gun at ONS

Oxifree Global is showcasing its revolutionary new Polymelt Service Gun SG1, for the first time to the biggest names in the oil and gas industry at Offshore Northern Seas ONS in Norway between 29th August and 1st September 2016.

The SG1 was carefully developed to allow asset owners a way of filling in small areas of Oxifree TM198 coating which have been removed for inspections, making maintenance in the field easier and more cost effective.

Corrosion and contamination in the harsh marine environment is a costly challenge for asset owners. Oxifree Global has developed Oxifree TM198, Oxitape and the new SG1 to tackle the global corrosion epidemic which ravages all sectors of the offshore industry.

"We are delighted to be showcasing the SG1 on our stand at ONS this year for the first time. After consultation and feedback with our clients and representatives, it was clear there was a need for a compact, lightweight solution for small coating touch ups," comments Ed Hall, Managing Director at Oxifree Global.

The SG1 allows end users to undertake inspections by removing small amounts of coating material to do visual inspections of the substrate, then simply reapplying material without the need to move around a large unit. Oxifree TM198 thermoplastic coating is a patent protected, organic, self-lubricating coating for the protection of metal components. It was developed by Oxifree Global to help reduce the cost of maintenance. TM198 lasts several years and has been proven to extend the lifecycle of components, reducing maintenance costs by at least 40%.

"One of the biggest challenges facing the oil and gas sector is how to increase asset lifetime and reduce the associated maintenance costs caused by corrosion and contamination," explains Ed. "Oxifree's unique TM198 coating and Oxitape offer unbeatable protection against corrosion. These, combined with the new SG1 – which can be operated by the end user and doesn't require a technician to be sent to site – will make inspections and maintenance in the field much easier and cost effective for asset owners."

The SG1 will be on display at Offshore Northern Seas from 29th August – 1st September on the Oxifree Global stand (Hall 6, booth number 6700. Visitors are invited to drop in to see the SG1 and see a demonstration of Oxitape and Oxifree's thermoplastic coating and to discuss the latest issues in corrosion technology with Oxifree's experts.

USA Oil and Gas Monitor
For Daily News Report and Analysis • www.usaoilandgasmonitor.com

September 2016 • Issue 9

BHP BILLITON to Cease Progression of Caroona Coal Project

BHP Billiton has agreed to cease progression of the Caroona Coal Project, through the cancellation of Exploration Licence EL 6505.
BHP Billiton Minerals Australia President, Mike Henry, acknowledged the NSW Government's willingness to come to an acceptable agreement in respect of the cancellation of EL6505.

"While we believe that Caroona would have been developed responsibly, we accept the Government's decision and appreciate its willingness to work with us to agree an acceptable financial outcome for the cancellation of our exploration licence," Mr Henry said.

"The Caroona Coal Project was studied extensively and developed cautiously for almost 10 years. We carried out extensive planning to ensure there would be no mining under the black soil plains, consistent with the conditions contained in our Exploration Licence.

"It was also subject to extensive scientific research which showed the proposed project could have been developed in an environmentally sustainable manner. We would like to express our sincere appreciation to the local community for working closely with us over the past 10 years through the project's lengthy consultation process."

WA Gas Reservation Policy Should Be Dumped, Not Expanded

Expanding Western Australia's gas reservation policy would destroy jobs and investment in regional communities, undermine the state's finances and reduce the volume of gas available for local use, the Australian Petroleum Production & Exploration Association APPEA has warned.

APPEA Chief Operating Officer – Western Australia Stedman Ellis said industry was deeply concerned that new National Party Leader Brendon Grylls had confirmed that he wanted to change the existing policy.

"The only change that should be contemplated is the complete removal of the policy," Mr Ellis said. 'Reservation is effectively a tax on gas production. It undermines energy security by discouraging the investment needed to increase gas supply and put downward pressure on prices. Gas producers are already struggling with low commodity prices. Imposing further costs on gas operations would undermine their international competitiveness and raise serious sovereign risk questions for the state. WA cannot afford to damage its hard-earned reputation as a reliable supplier of natural resources and a safe destination for global investment."

Mr Ellis urged the WA Nationals to focus on policies that encourage investment in gas production.

"New gas projects are good for regional, state and national economies. They provide jobs and investment, taxes and royalties and much-needed export revenues. Any move to expand the existing reservation policy by imposing further restrictions on gas exports would be extremely irresponsible. Forcing more gas into an already over-supplied domestic market in WA will also undermine investment in the onshore gas industry, which pays royalties directly to the State Government. This would actually mean less money to repair the state budget and less money for the National Party's Royalties for Regions program. It is akin to killing the goose that lays the golden eggs. There is a very good reason credible, independent bodies such as the WA Economic Regulation Authority, the ACCC and the Productivity Commission have all warned against gas reservation – the economic costs far outweigh any perceived benefits."

BakeHostetler- Three Reasons Why BOEM's Updated Financial Assurance and Risk Management Requirements are Unenforceable

The Interior Department's Bureau of Ocean Energy Management BOEM has finally issued its promised Notice to Lessees NTL No. 2016-N01 "Requiring Additional Security," which supersedes NTL No. 2008-N07 "Supplemental Bond Procedures." The new NTL, with multiple "linked" attachments, becomes effective on September 12, 2016. The new NTL largely concerns a lessee's ability to carry out its obligation to decommission wells, platforms and pipelines on oil and gas leases in all regions of the Outer Continental Shelf OCS. Lessees are already assessing how these new requirements will impact their OCS operations.

In many respects, the new NTL mirrors the Proposed Guidance advertised by the BOEM on September 22, 2015.1 despite inviting comment from stakeholders, the final "guidelines" fail to take into account several of the fundamental concerns expressed in comments submitted by companies and trade groups directly affected by the financial assurance requirements. As a result, the new NTL is subject to challenge on at least three bases.

First, the NTL's linked attachments conflict with the regulations. **Second**, by ignoring Asset Retirement Obligations ARO, which are already included in a company's audited financial statements, BOEM effectively double-counts the costs of removal by subtracting its own estimates of removal liability from a tangible net worth already reduced by AROs. **Third**, the 10 percent cap on self-insurance is a rule adopted outside of a proper rulemaking.

If the new NTL were found to be unlawful, the superseded Supplemental Bond Procedures in NTL No. 2008-N07 would not be automatically resurrected. At a minimum, it too violated the regulations in its disregard of the treatment of AROs in audited financial statements, contrary to 30 C.F.R. § 556.901(d)(1)(i), 81 Fed. Reg. 18112, 18170 March 30, 2016.

Reason Number One: The guidance contains language inconsistent with the regulations.

In one key respect, the new NTL appears as if it were written by two "committees" prohibited from communicating with one another. One committee, drafting the text of the NTL, recognizes that BOEM, before requiring a supplemental bond from a lessee,

September 2016 • Issue 9

has the burden of showing that the lessee lacks the financial capacity to perform its obligations under its OCS lease. The other committee, drafting the linked attachments, takes for granted that no one has the required capacity, making the only relevant question how big the supplemental bond must be. Lessees who are meeting with BOEM to discuss their financial capacity need to be mindful of the agency's inconsistent messaging. And lessees must be aware that the second message is unlawful.

The regulations applicable to leases require BOEM to ask two questions: First, does a lessee have the financial ability to carry out its present and future lease and regulatory obligations? 30 C.F.R. § 556.901(d). And second, if the answer is no, how much additional security must be provided? 30 C.F.R. § 556.901(e). BOEM recognizes this two-step process in the NTL No. 2016-N01: "The Regional Director will evaluate your financial ability to carry out your present and future obligations annually to determine whether you must provide additional security and, if so, how much additional security you must provide" and "if the BOEM Regional Director determines that the financial ability of any lessee or grantee for any lease, ROW, or RUE is not sufficient to assure performance of its lease, ROW, or RUE obligations, he or she may require the lessees or ROW and RUE owners to provide and maintain additional security" NTL page 2, I. and II. Emphasis added.

What happens if the answer is "Yes, a lessee has the financial ability to meet its financial obligations"? Pursuant to the regulations, this would be the end of the analysis. A company that meets the five criteria set forth in 30 C.F.R. § 556.901(d)(1) should not have to provide any additional security.

Despite its apparent recognition of the two-step process, the new NTL's incorporated attachments contradict this position. "This NTL discontinues the policies under NTL No. 2008-N07, whereby if BOEM determined that one or more co-lessees or co-owners had sufficient financial strength and reliability, it was not necessary to provide additional security." NTL page 1, Introduction. Or, as stated more explicitly in its summary of key changes to the NTL, "lessees will no longer be granted waivers, but may be eligible for self-insurance to meet some or all of their supplemental bond obligations."

So, BOEM's position appears to be that every company must provide supplemental financial assurance – it's just a matter of finding the right mix of self-insurance, if a company qualifies, and bonding through a tailored financial plan. But this is not merely a change in policy. It is contrary to the regulations. BOEM is not permitted by its regulations to assume that a company lacks adequate financial capacity; it must first affirmatively find the lessee does not.

Consider the following scenario: A major exploration and production company holds a 50 percent interest in a lease. The remaining 50 percent interest is split equally among three independent exploration and production companies. Major can demonstrate that it meets the NTL's thresholds for financial health:

- It exceeds the minimum thresholds for six of the nine "financial capacity" ratios set forth in the NTL.
- Its existing OCS lease production and proven reserves calculated using the PV-10 method are "significantly in excess" of its financial obligations.
- It has been operating in the OCS for more than five years.
- It has an S&P rating of BBB+ (an "investment grade" rating according to BOEM).
- It has a satisfactory record of compliance.

If BOEM properly applies the regulations, then Major should not be required to provide any additional financial assurance because the answer to the first question set forth in the regulations – "whether lessee has the financial ability to carry out your present and future obligations" – is answered in the affirmative.

What if the scenario is modified to concern a new lease issued to Major, as the 100 percent interest holder on a "sole liability property" – those leases for which there are no co-lessees and no prior interest holders? BOEM's guidance specifies that the agency will set a minimum credit rating below which BOEM would not allow the use of self-insurance on "sole liability properties." That minimum S&P credit rating is A-. Again, Major should not be required to provide any supplemental financial assurance for that property because it has met the five standards in 30 C.F.R. § 556.901(d)(1) for demonstrating financial ability to meet obligations. But will BOEM nevertheless require Major to provide a bond for the sole liability property because its credit rating is lower than A-?

Under either of these scenarios, BOEM's demand that Major provide supplemental financial assurance is unlawful.

U.S. Appeals Court Affirms RICO Judgment against Lawyer behind Fraudulent Ecuador Lawsuit

The United States Court of Appeals for the Second Circuit has unanimously affirmed a lower court decision, which found that the $9.5 billion judgment against Chevron Corporation in Ecuador was the product of fraud and racketeering activity, and unenforceable in the United States. The appeals court stated that there was "no basis for dismissal or reversal" of the district court's judgment, noting that "[t]he record in the present case reveals a parade of corrupt actions by the LAPs' legal team, including coercion, fraud, and bribery, culminating in the promise to Judge Zambrano of $500,000 from a judgment in favor of the LAPs."

In 2014, Steven Donziger, the American lawyer behind a fraudulent lawsuit against Chevron in Ecuador, was found by the U.S. District Court for the Southern District of New York to have violated the federal Racketeer Influence and Corrupt Organizations Act RICO, committing extortion, money laundering, wire fraud, Foreign Corrupt Practices Act violations, witness tampering and obstruction of justice in obtaining the Ecuadorian judgment and in trying to cover up the crimes committed by him and his associates.

"This decision, which is consistent with the findings of numerous judicial officers in the United States and South America, leaves no doubt that the Ecuadorian judgment against Chevron is the illegitimate and unenforceable product of misconduct," said R. Hewitt Pate, Chevron vice president and general counsel. "Chevron is pleased that the truth has prevailed over fraud and corruption."

During the seven-week RICO and fraud trial, Chevron presented unrebutted evidence detailing the extent of the fraudulent acts undertaken and directed by Donziger, his Ecuadorian legal team and other associates, including fabricating environmental evidence, pressuring scientific experts to falsify reports, plotting to intimidate judges into handing down favorable rulings, bribing court-appointed experts, ghostwriting court reports and even drafting the final judgment.

Chevron has never operated in Ecuador. Texaco Petroleum TexPet, which became a subsidiary of Chevron in 2001, was a minority partner in an oil-production consortium in Ecuador along with the state-owned oil company, Petroecuador, from 1964 to 1992. After TexPet turned its remaining share of the oil operations over to Petroecuador in 1992, pursuant to an agreement with Ecuador, TexPet agreed to conduct a remediation of selected production sites while Petroecuador remained responsible to perform any remaining cleanup. The government of Ecuador oversaw and certified the successful completion of TexPet's remediation and fully released TexPet from further environmental liability. Petroecuador, however, failed to conduct the cleanup it promised and has continued to operate and expand oil operations in the former concession over the past 20 years.

Since the extent of the fraud scheme was revealed, more than a dozen former insiders and allies have abandoned Donziger and his scheme, including his former co-counsel, environmental consultants, funders, investors, employees and Ecuadorian collaborators.

In May 2015, Brazil's Deputy Prosecutor General recommended to the country's Superior Court of Justice that the fraudulent Ecuadorian judgment not be recognized for enforcement, upholding both international and Brazilian law.

In December 2015, the Supreme Court of Gibraltar issued a judgment against Amazonia Recovery Ltd., a Gibraltar-based company set up by Donziger and his associates to receive and distribute funds resulting from the fraudulent Ecuadorian judgment. The court awarded Chevron$28 million in damages and issued a permanent injunction against Amazonia, preventing the company from assisting or supporting the case against Chevron in any way.

September 2016 • Issue 9

Stripper Wells Accounted for 11 per cent of U.S. Natural Gas Production in 2015

Stripper wells, also known as marginal wells, individually produce small volumes of natural gas or oil but in aggregate have provided 11 per cent to 15 per cent of total U.S. oil and natural gas production over the past decade. Natural gas stripper wells, so called because they are stripping the remaining natural gas out of the ground, are characterized as producing no more than 90,000 cubic feet per day over a 12-month period. EIA estimates that there were about 456,000 stripper gas wells in the United States operating at the end of 2015, compared with about 122,000 non-stripper gas wells.

These well counts include natural gas wells that may also produce some oil. Wells producing more than 6,000 cubic feet of natural gas per barrel of oil are considered gas wells, while wells producing 6,000 or less cubic feet of natural gas per barrel of oil are considered oil wells. Stripper wells contribute a small but significant portion of production of both natural gas and oil.

Stripper wells may have originally been high-volume wells, but through normal production declines now produce only small volumes. Because these wells usually have low ongoing maintenance costs, they are kept active and may continue to produce for many years, as long as they are economically feasible.

Despite each stripper well's small individual production, in aggregate they make a contribution to total natural gas production. The production share of stripper gas wells has remained relatively constant over the past 25 years, rising from about 10 per cent in 1991 to 15 per cent in 2006–09 and dropping again to about 11 per cent in 2015. The recent decrease in stripper wells' share of total production reflects the large increase in production from relatively prolific wells drilled in shale and tight gas formations with enhanced completion techniques. Because these wells, and non-stripper wells in general, produce at a much higher rate than stripper wells, they account for the bulk of total U.S. natural gas production.

CO2 Emissions from Natural Gas Surpass Coal as Fuel Use for the First Time in 44 Years

Energy-associated carbon dioxide CO2 emissions from natural gas are expected to surpass those from coal for the first time since 1972. Even though natural gas is less carbon-intensive than coal, increases in natural gas consumption and decreases in coal consumption in the past decade have resulted in natural gas-related CO2 emissions surpassing those from coal. EIA's latest Short-Term Energy Outlook projects energy-related CO2 emissions from natural gas to be 10 per cent greater than those from coal in 2016.

From 1990 to about 2005, consumption of coal and natural gas in the United States was relatively similar, but their emissions were different. Coal is more carbon-intensive than natural gas. The consumption of natural gas results in about 52 million metric tons of CO2 for every quadrillion British thermal units MMmtCO2/quad Btu, while coal's carbon intensity is about 95 MMmtCO2/quad Btu, or about 82 per cent higher than natural gas's carbon intensity. Because coal has a higher carbon intensity, even in a year when consumption of coal and natural gas were nearly equal, such as 2005, energy-related CO2 emissions from coal were about 84 per cent higher than those from natural gas.

In 2015, natural gas consumption was 81 per cent higher than coal consumption, and their emissions were nearly equal. Both fuels were associated with about 1.5 billion metric tons of energy-related CO2 emissions in the United States in 2015.

Annual carbon intensity rates in the United States have generally been decreasing since 2005. The U.S. total carbon intensity rate reflects the relative consumption of fuels and those fuels' relative carbon intensities. Petroleum, at about 65 MMmtCO2/quad Btu, is less carbon-intensive than coal but more carbon-intensive than natural gas. Petroleum accounts for a larger share of U.S. energy-related CO2 emissions because of its high levels of consumption.

Another contributing factor to lower carbon intensity is increased consumption of fuels that produce no carbon dioxide, such as nuclear-powered electricity and renewable energy. As these fuels make up a larger share of U.S. energy consumption, the U.S. average carbon intensity declines. Although use of natural gas and petroleum have increased in recent years, the decline in coal consumption and increase in non-fossil fuel consumption have lowered U.S. total carbon intensity from 60 MMmtCO2/quad Btu in 2005 to 54 MMmtCO2/quad Btu in 2015.

September 2016 • Issue 9

Homeostasis and Dynamic Equilibrium Keys to Navigating Change in a Volatile Oil and Gas Industry

Speaking at the just concluded Summer NAPE in Houston, Kathy Cleveland Bull explain to the industry CEOs and professionals on how they can use Homeostasis and Dynamic Equilibrium to Navigate Change. Below is the excerpt of the interview with USA Oil and Gas Monitor.

1. Tell us about yourself?

Kathy Cleveland Bull is a highly regarded professional speaker, trainer, author and consultant on organizational change. Professional highlights include:

• Keynoting an international medical conference in Rotterdam, The Netherlands inspiring professionals, caregivers and family members to successfully adapt to the diagnosis of dementia in a loved one
• Presenting a 2 day seminar on change for African business leaders in Nairobi, Kenya
• Appearing with "Dr. Phil" and Deepak Chopra at The Power Within Live Event touring cities in the US and Canada

• Co-authoring with Dr. Stephen Covey and others the book, Success Simplified – Simple Solutions Measurable Results
• Most recently, Kathy visited Dubai where she presented her successful change seminar to the Middle East executives of Ernst & Young Accounting Firm.

Kathy was one of just three professional speakers recommended by Dr. Spencer Johnson to provide training built upon his book Who Moved My Cheese? She has trained more than 250,000 people to successfully manage change in their work and life using the "cheese" metaphor. Prior to forming her own firm in 2001, Kathy was the Director of Training and Organizational Development at The Ohio State University. Over her 17 year career in higher education she held various leadership positions in student affairs administration, admissions, leadership development, student activities, and event planning at Bowling Green State University, North Carolina State University, and Ohio State.

A Phi Beta Kappa psychology graduate of Bowling Green State University, Kathy also holds two Master's degrees, in College Student Development and Guidance and Counseling. She has served her alma mater as a member of the university's Alumni Board of Trustees and chaired the Arts and Sciences Board of Advocates. She received the Outstanding Recent Graduate Award from the BGSU Alumni Association and was honored to deliver the May 2006 Commencement Address at BGSU.

2. In your speech at the NAPE — you mentioned Homeostasis and dynamic equilibrium as key to helping the industry navigate change. Can you explaining these terms and why you think it's so important?

As I stated in my remarks we must embrace a new paradigm for thinking about change which includes staying true to our core values and organizational mission while vigorously embarking on a journey

September 2016 • Issue 9

of transformation in all other endeavors. This new paradigm I am offering includes both the homeostatic model - where energy is only exerted to counteract change, to bring the system back to a static balance and the dynamic equilibrium model - where the energetic tension of opposing forces is itself the balance, and in response to change, a new organizational balance may emerge. This type of balance is more appropriate for stimulating progress, when the innovative response is needed to effectively adapt to an ever changing global marketplace. In the homeostatic model change is the enemy of balance. It is to be avoided or counteracted. In business, I would argue that homeostatic balance may be an appropriate model for preserving the core. Because threats must be avoided or counteracted so that the organization's essential integrity remains steadfast and protected. So, we need both, homeostasis in order to preserve our core, and dynamic equilibrium in order to stimulate progress.

3. Talking about adapting to change- which for the industry is the new low price environment, what is the blue print/roadmap to successfully adapting to change?

I borrow the road map from the work of William Bridges and his Three Phases of Change. In the first phase, which is called Endings, we leave the good old days when oil was $100 a barrel behind and begin to face head-on the reality of today. It can be frightening and emotional as it is characterized by grieving of the previous reality we knew. The second phase is called The Neutral Zone by Bridges where we are surrounded by ambiguity. We recognize that the old business environment is gone, but we don't know what the future will look like. We have more questions than answers and for those who "take charge of change" it can be a very fruitful time in business as the sense of urgency thrusts us towards new possibilities. Finally, we find ourselves in a New Beginning, phase three. Here we celebrate our new reality and recognize that those who remain in business have successfully navigated the white waters of global change.

4. What is your takeaway from this year's summer NAPE conference?

The oil and gas industry is poised to transform the way it does business through technological advances and innovative business solutions. This can only happen if leaders and professionals in the industry embrace the change they face and engage in a "new culture of inquiry".

About NCompass Consulting

Kathy Cleveland Bull founded her company, NCompass Consulting, in 2001 and since that time has helped clients on four continents "Navigate the Art and Science of Change." NCompass Consulting provides keynote speeches and professional development seminars for Fortune 500 companies, educational institutions, government agencies, not-for-profit organizations and religious institutions. In addition, we provide consulting service solutions for successful change strategies, human performance, organizational culture change and strategic planning. Her recent clients include such diverse organizations as The Social Security Administration, The US Department of Energy, Nationwide Financial, Pfizer, Novartis, Ernst and Young, and more than 50 colleges and universities in the US and Canada.

More Tips for Oil and Gas Executives on How to Navigate Change from - Kathy Cleveland Bull

What's the No. 1 mistake executives make when facing an industry downturn or organizational threat?

The number one mistake executives make is reacting instead of responding. When we experience a threat, we have a natural, hard-wired tendency towards defensive reaction. That tendency to react is based on our personal history of other threats we have experienced in life and may or may not be appropriate to the current situation. So what we need to do is to learn to hit the pause button and respond to the threat from an informed and grounded place.

What is your advice for professionals who have experienced a significant change or setback in their career?

Those who have been in the workplace for decades have had to learn to deal with change, in fact a seasoned leader's success has likely been measured by how well she or he has navigated change. However, for some younger executives, successful change management isn't yet fully integrated into their skill set. To a younger professional a setback may come as a surprise and may instinctively be viewed as a loss. What's important is to integrate change into our way of thinking about and seeing the world. Understand that how you respond to the change will determine whether that change becomes a loss or an opportunity.

How can leaders avoid change in their industry and company?

One of the hallmarks of a good leader is viewing change as inevitable, positive and necessary for corporate success. In business and in life, the key is to keep your mission and values constant while being open to change in every other way. True success in business is not a static result but a dynamic approach to living with constant change and the ebb and flow of all the different aspects of organizational life. Not judging as good or bad, fortunate or not, but just part of the business reality to meet head on. This approach isn't destabilized by change – but includes change as a reality.

What tips would you share with executives who are facing change in their organization or industry?

Number one, understand that we always have the power of choice in changing times. We may not have any control or influence over the direction of change, but we always have the power to choose our own response, and therein lies the key to our success. Number 2, be open to new ideas and proactively scan the horizon for shifts, trends and possibilities. And number 3, understand that most people resist change because of a loss of control, so invite the largest possible participant group to be the architects of change.

Seven Regions in U. S. Accounts For 92 Per Cent of Domestic Oil and Gas Production

The United States Drilling Productivity Report which uses recent data on the total number of drilling rigs in operation along with estimates of drilling productivity and estimated changes in production from existing oil and natural gas wells to provide estimated changes in oil and natural gas production for seven key regions. These seven regions account for 92 per cent of domestic oil production growth and all domestic natural gas production growth during 2011-14.

While shale resources and production are found in many U.S. regions, EIA Drilling Productivity Report focuses on the seven most prolific areas, which are located in the Lower 48 states.

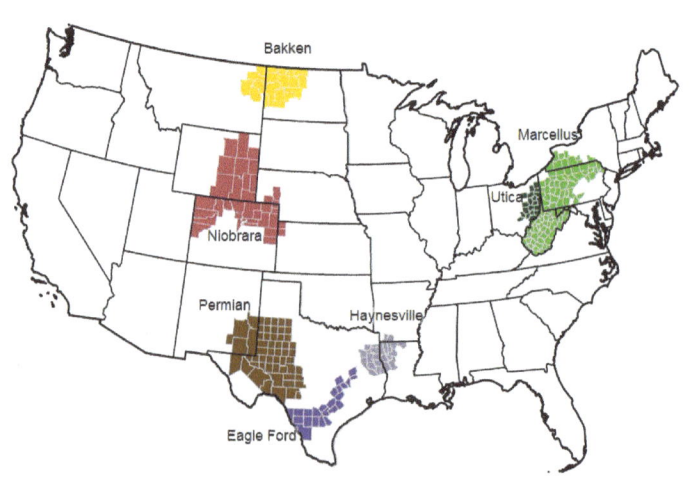

Region	Oil production thousand barrels/day			Gas production million cubic feet/day		
	August 2016	September 2016	change	August 2016	September 2016	change
Bakken	968	942	(26)	1,570	1,543	(27)
Eagle Ford	1,079	1,026	(53)	5,806	5,594	(212)
Haynesville	46	45	(1)	5,880	5,844	(36)
Marcellus	38	37	(1)	17,843	17,810	(33)
Niobrara	377	370	(7)	4,182	4,113	(69)
Permian	1,974	1,977	3	6,868	6,863	(5)
Utica	73	73	-	3,674	3,683	9
Total	4,555	4,470	(85)	45,823	45,450	(373)

September 2016 • Issue 9

USA Oil and Gas Monitor
A RGT Media Communications Corp.

SUBSCRIPTION FORM

First Name _____ Middle _____ Last _____

Current Job Title _____ Job Title Code _____

Company Name _____

Preferred Mailing Address - (Circle One)

   Business   Residence

Street _____ (No PO Boxes Please)

City _____ State _____ Zip _____

Country _____

Day Phone _____ If outside U.S., include country code. (ex: 000-000-000-0000)

Fax _____ Email _____

Form Instructions:

Email completed form to subscribe@usaoilandgasmonitor.com or mail form with check to the address below.

RGT Media Communications Corp.
10777 Westheimer Road #1100
Houston Texas 77042

1 Year Digital Subscription

For non-Texas subscribers - $119.88

Subscribers living in Texas – pays $119.88 plus 8.25% state tax $9.89 = $129.77

1 Year Print Subscription

For non-Texas subscribers - $144
Please add shipping cost and multiply by 12 (for example $1.67 x12) = 20.04
Subscribers living in Texas – pays $144 plus 8.25% state tax $11.88= $155.88
Please add shipping cost and multiply by 12 (for example $1.67 x12) = 20.04

Shipping Cost (calculated by weight)

Circle choice from the following option and add to the subscription cost

First Class 1- 5 business/days = $1.67

Fedex Shipping 1-3 business/days= $6.40

USPS Priority 1-3 business/days= $3.56

International First Class 1-7 business/days= $12.44

You can also pay for subscription online by visiting our website:
www.usaoilandgasmonitor.com/subscribe
Wire transfer, call Jewel Spring, 832-486-0095 for any questions.

Payment Method

Card Type (circle one)

Amex   Visa   Master   Discovery

Card No.

_____

Expiration Date

_____

CSV No.

_____

Name on Card

_____

By Check

Check No.

_____

The 30 Largest U.S. Refineries with Capacity Over 100,000 Barrels per Atmospheric Crude Oil Distillation

Companies	Barrels per Calendar Day- Jan 1,2016
VALERO ENERGY CORP	2,062,300
EXXON MOBIL CORP	1,857,500
MARATHON PETROLEUM CORP	1,794,000
PHILLIPS 66 COMPANY	1,612,200
MOTIVA ENTERPRISES LLC- 50% Royal Dutch/Shell Group, 50% Saudi Aramco	1,075,700
CHEVRON CORP	951,271
TESORO CORP	831,030
PDV AMERICA INC- CITIGO	761,240
PBF ENERGY CO LLC	694,700
BP PLC	651,000
KOCH INDUSTRIES INC	585,630
WRB REFINING LP- 50 percent Phillips 66, 50 percent Cenovus	482,000
HOLLYFRONTIER CORP	467,350
ROYAL DUTCH/SHELL GROUP	437,975
CARLYLE GROUP	335,000
DEER PARK REFINING LTD PTNRSHP - 50% Royal Dutch/Shell Group, 50% Pemex	285,500
ACCESS INDUSTRIES	263,776
ALON ISRAEL OIL COMPANY LTD	237,500
TOTAL SA	225,500
DELTA AIR LINES INC	190,000
CVR ENERGY	185,000
CALUMET SPECIALTY PRODUCTS PARTNERS, L.P	169,920
DELEK GROUP LTD	155,000
BP-HUSKY REFINING LLC - 50 percent BP, 50 percent Husky	153,000
HUSKY ENERGY INC	152,000
WESTERN REFINING INC	147,500
CHS INC	145,600
PETROBRAS AMERICA INC	112,229
SINCLAIR OIL CORP	109,500
SUNCOR ENERGY INC	103,000
Total	**17,233,921**

September 2016 • Issue 9

A Total of 141 U.S Operable Refineries as of January 1, 2016

Years	U.S. Number of Operable Refineries as of January 1, 2016
1982	301
1983	258
1984	247
1985	223
1986	216
1987	219
1988	213
1989	204
1990	205
1991	202
1992	199
1993	187
1994	179
1995	175
1996	
1997	164
1998	
1999	159
2000	158
2001	155
2002	153
2003	149
2004	149
2005	148
2006	149
2007	149
2008	150
2009	150
2010	148
2011	148
2012	144
2013	143
2014	142
2015	140
2016	141

Key term = Definition

Atmospheric Crude Oil Distillation- The refining process of separating crude oil components at atmospheric pressure by heating to temperatures of about 600º to 750º F (depending on the nature of the crude oil and desired products) and subsequent condensing of the fractions by cooling.

Barrels Per Calendar Day- The amount of input that a distillation facility can process under usual operating conditions. The amount is expressed in terms of capacity during a 24-hour period and reduces the maximum processing capability of all units at the facility under continuous operation to account for the following limitations that may delay, interrupt, or slow down production:

- the capability of downstream facilities to absorb the output of crude oil processing facilities of a given refinery. No reduction is made when a planned distribution of intermediate streams through other than downstream facilities is part of a refinery's normal operation;
- the types and grades of inputs to be processed;
- the environmental constraints associated with refinery operations;
- the reduction of capacity for scheduled downtime due to such conditions as routine inspection, maintenance, repairs, and turnaround; and
- the reduction of capacity for unscheduled downtime due to such conditions as mechanical problems, repairs, and slowdowns.

September 2016 • Issue 9

North American refiners enjoyed a large discount to global crude oil prices for several years, measured by the difference between the U.S. composite refiner acquisition cost and North Sea Brent crude oil prices. With price discounts often in the double digits, North American refiners were consistently more profitable than global and European refiners, which on average paid higher prices for crude oil. Since

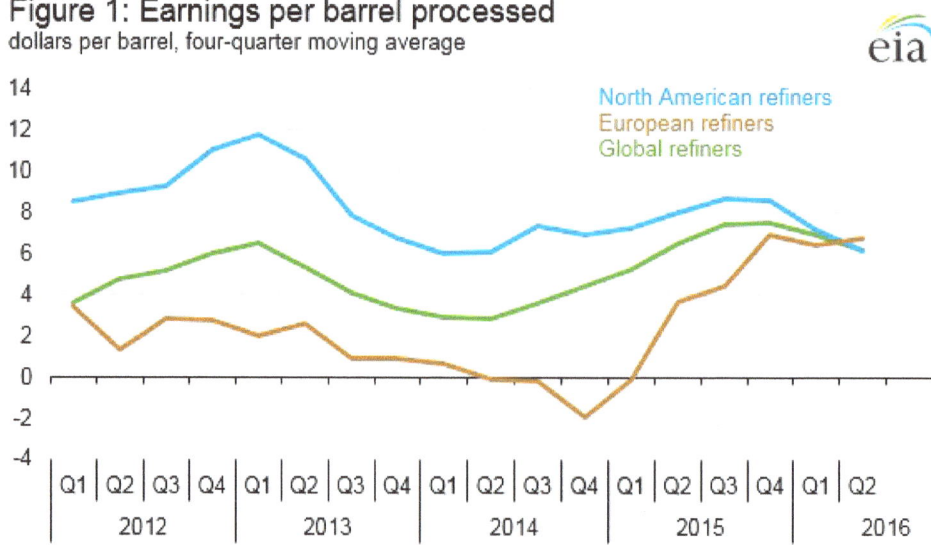

Figure 1: Earnings per barrel processed
dollars per barrel, four-quarter moving average

North American refiners
European refiners
Global refiners

Sources: U.S. Energy Information Administration, based on Evaluate Energy.

North American refiners— Less Profitable than European Refiners for the Second Quarter 2016

the third quarter of 2015, the discount has not widened beyond $4.50 per barrel, reducing some U.S. refiners' competitive advantage on costs.

Refinery earnings were lower in second-quarter 2016 compared with the same time last year and are converging among different locations globally. Lower crack spreads, the price difference between crude oil and petroleum products, contributed to declining profits for some refiners compared with 2015. Changes in North American and European crude oil price differentials are likely contributing to the convergence in profits.

Recently released second-quarter statements from 27 companies show that 21 experienced a year-over-year decline in 2016 in refining profits, as measured by earnings per barrel processed. The decline in earnings was commensurate with the decline in crack spreads in the second quarter. In addition to changes in crack spreads, which serve as an indicator of refinery profits, earnings per barrel account for other costs, such

as transportation costs and other operating expenses. Also, each refiner uses different crude oil blends and produces different yields of refined products, which will show differences among refiners in their per barrel earnings. Analyzing the group of companies by primary refining region shows refinery earnings converging over the past year. The group of North American refiners consist of 14 companies with operations mainly in the United States and Canada, while seven companies constitute the European group and six the Global group, so-named because its companies' operations are geographically diversified and affected by crude oil and petroleum product prices in many regions of the world. Companies in North America and Europe, on the other hand, may be more subject to differences in local markets.

One factor contributing to a convergence in refinery profits is an increase in the U.S. average refiner crude oil acquisition cost compared with global refiner acquisitions costs. Because crude oil and petroleum product prices are the two largest factors that affect a refiner's profits, changes in the cost of crude oil acquisition can have a significant effect on profitability.

In the European market, refiners may be achieving increased efficiency through consolidating

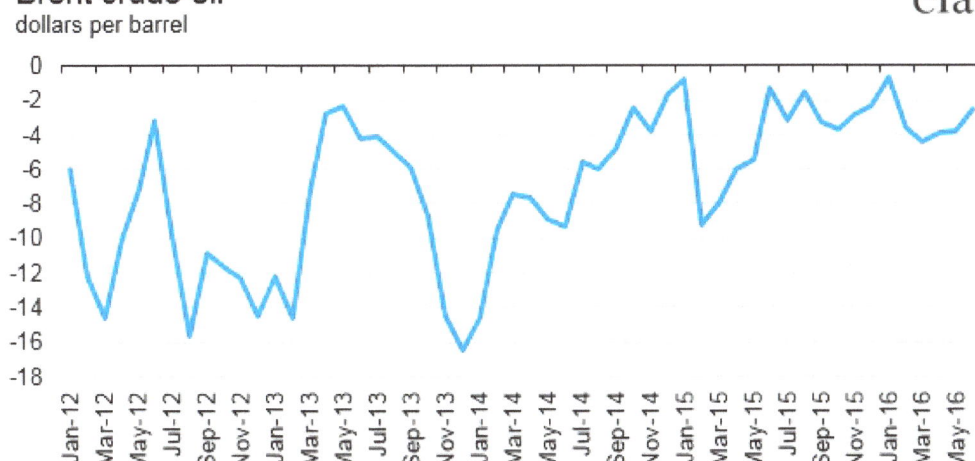

Figure 2: U.S. composite refiner acquisition cost minus Brent crude oil
dollars per barrel

Sources: U.S. Energy Information Administration, *Petroleum Marketing Monthly*; Thomson Reuters.

oils, giving some U.S. refiners a cost advantage. These factors began to reverse in 2014 and 2015. New pipeline infrastructure increased takeaway capacity to refining areas, such as the BridgeTex and Cactus pipelines in West Texas and Flanagan South in the Midwest. U.S. crude oil production declines, which began on a year-over-year basis in December 2015, also contributed to a comparatively tighter crude oil market in North America.

operations. The European companies in this analysis reduced distillation capacity 248,000 barrels per day in 2015, the fourth consecutive year of reductions. In addition, the crude oil market in Europe is comparatively looser than in North America, as Russian, Iranian, and Iraqi crude oil production has increased. Some refiners may be receiving lower costs for crude oil than other refiners as oil producers compete to maintain market share, which increases refinery profits. For example, the discount of Mediterranean Urals—a Russian crude oil many inland European refiners process—averaged greater than $2.00/b for most of 2016, which may have contributed to higher profits in recent quarters.

Many of the factors that contributed to the wide spread were driven by rapid increases in U.S. and Canadian crude oil production that were not met with increases in infrastructure to allow the oil to be moved to refining centers inexpensively. Crude oil producers typically received a price lower than global prices for similar quality crude

Crack spreads are lower in the third quarter and if the smaller spreads continue, it suggests third-quarter refinery profits will be lower globally. Absent meaningful changes in geographic crude oil and petroleum product price differentials, however, refinery profits will likely display smaller variability across different locations.

Figure 3: Mediterranean Urals crude oil price minus Brent crude oil
dollars per barrel, 20-day moving average

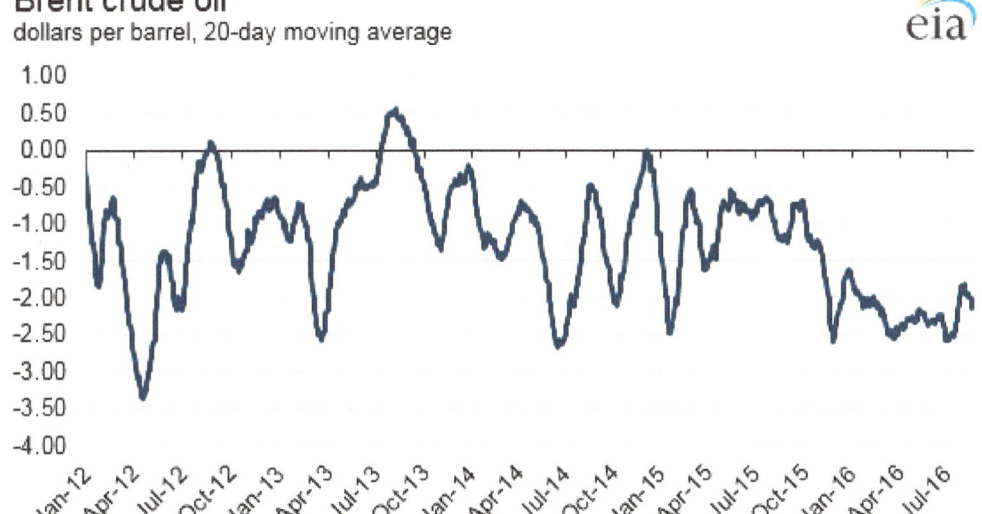

Source: U.S. Energy Information Administration, based on Thomson Reuters.

September 2016 • Issue 9

OriginClear's licensee- ECT Services and Solutions Unveiled a New Wastewater Treatment System

- Oil & gas operators now have a complete wastewater recycling solution--and can share water with the agriculture industry
- Farmers have the opportunity to use new, diverse water sources and dramatically cut down their water costs
- Both industries can meet increasingly strict federal and state water use regulations

OriginClear Inc. OTC/QB: OCLN, a leading provider of water treatment solutions, has announced that an OriginClear licensee will unveil a complete, modular system "ECOPOD" – now in operation at an oil field near Bakersfield, California – to clean up produced water. The ECOPOD pilot scale system uses OriginClear's Electro Water Separation™ EWS, requires no chemicals and reliably processes 340 barrels per day of oilfield produced water with minimal operator supervision.

Talbott Howard, CEO and Founder of OriginClear licensee ECT Services & Solutions ECT, will unveil ECOPOD System 1.0 at today's "Tackling the Drought: Exploring Safe, Innovative Water Sources" conference at California State University at Bakersfield CSUB, sponsored by the California Independent Petroleum Association and the California Energy Research Center.

Mr. Howard will show how ECOPOD System 1.0 can successfully and competitively treat produced water and recycle it into irrigation quality water for local beneficial reuse.

"Our own Dr. Cabrales of the Department of Physics and Engineering began working with OriginClear to research and validate its technology for oil and gas water cleanup," said Robert M. Negrini, Ph.D., Director of CSUB California Energy Research Center School of Natural Sciences, Mathematics, and Engineering. "We are eager to see how it all came together in the field."

ECT developed the compact ECOPOD System 1.0 to transform produced water into clean water for use in irrigation and potable water applications. The centerpiece is OriginClear's commercial-scale P3000, designed to treat up to 3000 barrels per day. The ECT team implemented additional scrubbing and polishing steps to achieve potable grade water. The entire process is chemical-free and is extremely energy-efficient while reducing the rate water evaporates during re-use.

"Drilling yields up to 13 gallons of wastewater for every gallon of petroleum," said Talbott Howard. "Federal and state pressure on the oil industry to recycle that water – potentially increased byProposition 1 spending – is creating a major business opportunity for ECT to treat up to 42 billion gallons a year in Kern County. There's also a possible bonus in recovering more valuable crude oil and rare earth minerals. Prospective customers are now visiting the site and we are ready for commercial deployment on a Just-In-Time schedule."

"My hat is off to the ECT team for succeeding under intense deadlines," said Jean-Louis JL Kindler, President of OriginClear's Technology division. "I especially want to thank Nicholas Eckelberry and Michael Green who have spent many hours on this project."

5 Ways Modern Integrated Automation Makes Plants Safer- Siemens Solution

By upgrading their automation technology, manufacturing plants are able to integrate safety functionality into all standard components for improved system performance and productivity. Learn five ways today's most advanced integrated automation technology helps plants exceed global safety-compliance requirements quickly and cost-effectively.

1. Integrated Safety Functionality In the past, plant engineers had to hardwire e-stops, gate interlocks, light curtains and other monitoring and safety-shutdown equipment as "bolt-on" accessories to a separate safety system. With today's most innovative automation technology, however, safety features are seamlessly integrated into all of the components. All programming — whether it's for safety or standard hardware — is done within the same software package, so safety planning is standardized throughout the system. In addition to making plants safer, integrated safety reduces total cost of ownership, and places less strain on engineering and maintenance personnel. It also enables greater system availability — due to improved diagnostics and troubleshooting — and greater operational flexibility, as plants can reconfigure their floor layouts and machine placements more easily.

2. Compliance with Safety Standards Unlike older or underperforming systems, today's modern, high performing automation components maximize process safety by meeting the latest international standards for fault tolerant applications. These standards, including IEC 61508, ANSI/ISA-84, IEC 62061, EN ISO 13849-1 and IEC 61511, cover the planning, documentation and assessment of all activities required to manage safety throughout the entire life of a system. IEC 61508, for example, is an international standard for the functional safety of automation components that are designed to detect potentially dangerous conditions and initiate corrective or preventative action. The standard establishes criteria for a Safety Integrity Level SIL, which describes a safety function's probability of a dangerous failure per hour.

3. Transmission of Safety-Related Data With advancements in automation technology and the emergence of networked safety, it's no longer necessary to run two separate fieldbuses for safety and non-safety data. Plants can use a standard fieldbus to transmit safety-relevant data, which reduces wiring complexity, system costs and training demands while improving diagnostic capabilities and freeing up space in the control cabinet. The emergence of PROFIsafe

September 2016 • Issue 9

— an integrated safety profile developed by the global consortium PROFI International — extends the standard communications protocol to address special requirements necessary to conform to standards such as IEC 61508. For example, PROFIsafe adds elements such as message numbering and dataconsistency checks to rule out typical network messaging faults, enabling networked safety devices to meet the reliability requirements of SIL 3 as prescribed by IEC 61508.

4. Deeper Visibility into Problems Advanced diagnostic capabilities provide deeper, real-time visibility into system performance and behavior, enabling plants to be more proactive when addressing potential problems. With integrated safety, it's no longer necessary to constantly interrogate the system to determine if e-stops and other I/O safety devices are functioning properly. Today's PLC systems conduct those validation tests automatically and report the results to the controller. Since the controller doesn't have to initiate and send the commands across the network to conduct validation tests, the process consumes less code and less bandwidth, while making the entire system more efficient and less vulnerable to programming errors. With safety solutions integrated directly into standard control architectures, plants

can leverage automation technology to address two separate issues: functional safety and system availability. Integrated safety helps to minimize accidents and downtime by enabling operators to diagnose hazardous conditions more intelligently and quickly.

5. Remote Diagnostic Capabilities Today, modular components such as PLCs, HMIs, drives and network switches offer integrated diagnostic functions, which makes system monitoring, troubleshooting, and maintenance easier and safer than ever before. With the integration of wireless technology, plant personnel can view the status information of all components from a networked computer or mobile device. The system can send automatic alerts to the mobile devices of responsible parties, who can securely log into the system, if necessary, to assess and correct the situation. With real-time remote diagnostic capabilities, operators and maintenance technicians are empowered to detect, report, and clear faults quickly and safely. Technicians, for example, can troubleshoot issues inside motor-control cabinets from a safe distance, minimizing the need to wear the specialized personal protective equipment at all times that may be necessary to shield them from arc-flash hazards.

EG Ronda 2016 Moves to Singapore for 3rd Stage of Oil and Gas Licensing Round Promotion

The Equatorial Guinea oil and gas licensing round, EG Ronda 2016, will come to Singapore in September as the Ministry of Mines and Hydrocarbons pushes forward on promoting available blocks. The visit to Singapore will open opportunities for companies in South-East Asia and Australia to take part in the round. H.E. Gabriel Mbaga Obiang Lima, Minister of Mines and Hydrocarbons, said Singapore's position as a leading energy center in South-East Asia makes it the ultimate venue for EG Ronda.

"We have considerable admiration for what Singapore has accomplished in becoming an important global energy hub," said H.E the Minister. "It is a model for Equatorial Guinea's long-term development and a reference point for oil and gas trading in South-East Asia. We look forward to having excellent discussions with organizations in the region."

During the Singapore visit, the Minister will make a keynote presentation during the CWC LNG & Gas Series: 8th Asia Pacific Summit, to be held September 20-23 at the Grand Hyatt Hotel. The conference will

bring together key stakeholders in the gas and LNG supply chain. Equatorial Guinea has a large portfolio of gas projects in operation or under development, including EG LNG and the pioneering Fortuna FLNG development which would be the first of its kind in Africa.

Tanya Crossick, Managing Director of CWC, stated: "CWC Group is delighted to welcome H.E. Gabriel Mbaga Obiang Lima, Minister of Mines and Hydrocarbons of Equatorial Guinea, at the 8th World LNG & Gas Series: Asia Pacific Summit. As an important supplier of gas, we look forward to stimulating discussion around new and existing gas supply from Equatorial Guinea and the opportunities that this brings to the global LNG & Gas sector."

EG Ronda 2016 launched on June 6, 2016 at the Africa Oil & Power conference and will conclude on November 30, 2016. The MMH invites bids on all open acreage outside of blocks under direct negotiation.

Licensing Rounds - Western Australian Petroleum Acreage Release Round 1 Of 2016

LICENSING ROUNDS

WESTERN AUSTRALIAN PETROLEUM ACREAGE RELEASE Round 1 of 2016

Proposed release to be announced at DMP Petroleum Open Day, and online from 13 September 2016.

Onshore Canning Basin – Lennard Shelf and platform areas

- Interest in the onshore Canning Basin has revived significantly in recent years, with new play oil discoveries at Ungani 1 and Ungani Far West 1 (both by Buru Energy), good oil and gas shows in Ordovician shale in the unconventional Theia 1 well, and large estimates for shale and tight gas.
- Lennard Shelf fields have produced oil from Devonian carbonates and Permian-Carboniferous sandstones.
- Numerous oil and gas shows have been encountered in the vicinity of the platform area, which may be prospective for sub-salt Ordovician plays.

Onshore Officer Basin

- The basin resembles Neoproterozoic successions in Oman and Russia that contain giant oil and gas fields.
- Numerous oil and gas shows have been encountered in Western Australian and South Australian mineral, petroleum and stratigraphic drillholes.
- Sub-salt and unconventional hydrocarbons maybe present.

(work program bids close 9 March 2017).
Information goes live on release date 13 September 2016 - visit www.dmp.wa.gov.au/acreage_release

Release information includes prospectivity of release areas and of relevant basins, available data listings, new bid assessment guidelines, land access and environment considerations, schedule of fees, and how to make a valid application for an Exploration Permit.

This information will also be available on USBs for distribution at events and meetings. Updated work program bid assessment criteria include applicant profile, geological evaluation and exploration rationale, environmental management, native title and heritage management, and land access management.

WESTERN AUSTRALIAN GEOTHERMAL ACREAGE

Companies may apply at any time via a Geothermal Special Prospecting Authority with an Acreage Option.
www.dmp.wa.gov.au/Petroleum/Geothermal-acreage-releases-1590.aspx

FURTHER INFORMATION ON STATE ACREAGE RELEASES

Should you require any further information or assistance, please contact DMP's Petroleum Division or the Geological Survey of Western Australia. All enquiries will be dealt with in strictest confidence.

COMMONWEALTH (Federal) OFFSHORE ACREAGE RELEASE

Refer to Department of Industry (DOI).
http://www.industry.gov.au/resource/UpstreamPetroleum/OffshorePetroleumExplorationinAustralia/Pages/AnnualReleaseofAustralianOffshorePetroleumExplorationAcreage.aspx

"OVER THE COUNTER" ONSHORE ACREAGE ACQUISITION

Special Prospecting Authority with an Acreage Option (SPA/AO)
www.dmp.wa.gov.au/Petroleum/Understanding-Petroleum-Titles-4224.aspx

POTENTIAL FARM-IN OPPORTUNITIES

- For basin maps showing companies and petroleum titles onshore and State Waters, Western Australia refer to: www.dmp.wa.gov.au/petop "Petroleum titles in Western Australia showing operators and applicants". These maps should be updated by mid-August 2016. Some of the companies shown may be interested in capital to expand exploration activities. Many of these companies have their own websites which can be searched.
- Disclaimer: due diligence is required; the map information is believed to be correct at the time of publication, but no responsibility is taken for its accuracy, completeness, or that any or all of the companies are interested in farm-in deals.

MERGER AND ACQUISITION

Some of the participants in the State are junior companies and there may be scope for acquiring acreage by means of company takeover, subject to Foreign Investment Review Board guidelines.
www.firb.gov.au

September 2016 • Issue 9

LEUM TITLES IN WESTERN AUSTRALIA
NG OPERATORS AND APPLICANTS — AUGUST 2016

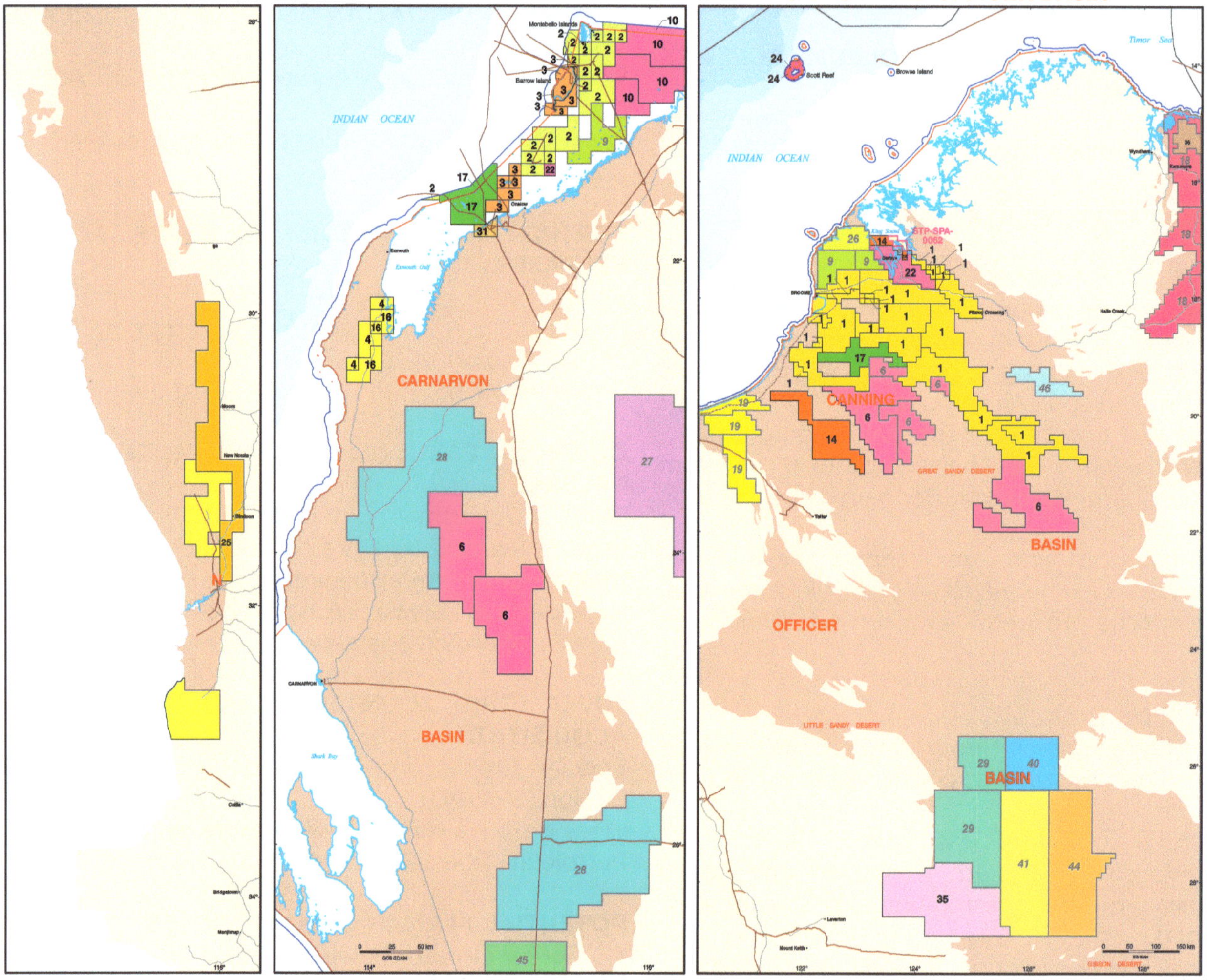

CARNARVON BASIN

CANNING BASIN AND OFFICER BASIN

CARNARVON BASIN

Operators
2	Quadrant
3	Chevron Australia
4	Empire Oil *
16	New Standard Onshore
10	Camarvon Petroleum
16	Rough Range Oil *
17	Finder No 3
22	Oil Basins
31	DBP Development

Applicants
8	Goshawk Energy
27	Pangaea Resources
28	Rusa Resources
45	Palatine

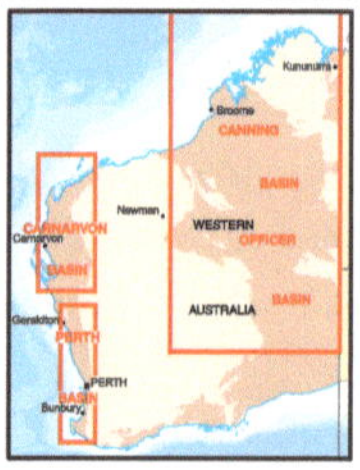

Note:
* # + x Symbols indicate associated companies

SPA 1 AO Special Prospecting Authority with Acreage Option

Basement Rock

Sedimentary Basin

Outer Limit of WA Coastal Waters (AMB 2014)
Territorial Baseline (AMB 2014)
Road
Pipeline Licence
Proposed Pipeline

CANNING BASIN AND OFFICER BASIN

Operators
1	Buru Energy
5	AWE Perth
6	New Standard Onshore
14	Gulliver Productions #
17	Finder Shale
22	Oil Basins
24	Woodside Energy
35	Officer Petroleum
36	Onshore Energy

Applicants
8	New Standard Onshore
9	Goshawk Energy
16	Canning Petroleum
19	Oilex
26	Black Rock Mining
29	Tamboran Resources
40	Australasian Energy
41	Australian Petroleum Portfolio
44	Liberty Petroleum
46	Strata-X

This map was produced using information made available to the Department of Mines and Petroleum (DMP) from various sources. DMP and the State cannot guarantee the accuracy, currency or completeness of the information. DMP and the State accept no responsibility and disclaim all liability for any loss, damage or costs incurred as a result of any use of or reliance whether wholly or in part upon the information provided in this publication or incorporated into it by reference. A record of title operators and proposed title operators is kept for administration purposes only and has no basis in legislation.

SIZE COMPARISON MAP
United Kingdom
Alberta
Texas
AUGUST 2016

s

of the above-mentioned companies that could do with further capital to expand s.

government would like to see accelerated exploration, production and e.

	Web address
ap	www.dmp.wa.gov.au/petop
ister (titles system)	www.dmp.wa.gov.au/pgr
	www.dmp.wa.gov.au/Documents/Petroleum/PD-RES-PUB-100D.pdf
	www.dmp.wa.gov.au/acreage_release
	www.dmp.wa.gov.au/wapims
ries	www.dmp.wa.gov.au/Petroleum/Major-prospective-areas-1584.aspx
	www.dmp.wa.gov.au/Petroleum/Maps-and-geospatial-information-1606.aspx

Australia Petroleum

www.dmp.wa.gov.au

TITLES REGISTER
For the on-line Petroleum and Geothermal Register (PGR) refer to:
www.dmp.wa.gov/pgr

PETROLEUM DATA ACCESSIBLE AND INEXPENSIVE
Internet downloads free.
Refer to WAPIMS:
www.dmp.wa.gov.au/Petroleum-and-Geothermal-1497.aspx

DMP PUBLICATIONS
Explorer's Guide Petroleum and Geothermal Energy Western Australia 2014
Comprehensive guide to prospectivity, geothermal, carbon capture and storage, access to data, and administration, including land access, environmental assessment, finances, and safety. www.dmp.wa.gov.au/Documents/Petroleum/PD-RES-PUB-100D.pdf

UPCOMING EVENTS
Talk to staff from the Western Australian Government at the following events:

Good Oil Conference, Hyatt Hotel, Perth, Western Australia, 13-14 September 2016
For further information about this event go to www.riuconferences.com.au

DMP Petroleum Open Day, Hyatt Hotel, Perth, Western Australia, 15 September 2016
Presentations planned include trends in petroleum, acreage release, Canning Basin, innovation, resource guidelines, and safety in petroleum, with opening remarks by the new Minister for Mines and Petroleum, Hon Sean L'Estrange MLA. There will be an extensive poster display, demonstrations of PGR and WAPIMS, an environment and safety workshop, and a panel discussion.

For further information about this event go to www.dmp.wa.gov.au/Petroleum/Petroleum-Open-Day-2016-18795.aspx

PETEX, London, UK, 15-17 November 2016.
For further information about this event go to www.petex.info/

September 2016 • Issue 9

Summer NAPE 2016- Takeaways

Summer NAPE 2016 saw nearly 3,000 upstream dealmakers, financers and energy professionals in Houston. Over two full days, insightful Business Conference presentations, packed Prospect Presentations and buzzing activity on the Exhibit Floor proved that NAPE continues to be the place for oil and gas industry decision makers to be seen and do business in all market conditions.

The expo provided opportunities for attendees to reconnect and forge valuable new contacts and partnerships. Below are some of the takeaways from participants and exhibitors.

Chad Barbe, Landman with Manzano Energy Partners II, LLC said, "Our presence at Summer NAPE 2016 and as a Prospect Presenter has opened the door for several new business opportunities for us," said.

Bill Britain, President / Chairman of EnergyNet said, "We're busier than we've ever been,". It's so great, NAPE. We've had an opportunity to really talk to people and have quality conversations about their properties and their business. We love it here and would not miss it."

"Western Land Services is a life-timer; we'll always come to NAPE. Even when the market is slow, the value of seeing new clients and having a presence here is invaluable," said a representative with the company.

Arthur Medina, Vice President with Purple Land Management said, "Summer NAPE gave us the opportunity to connect with our key industry contacts and colleagues,".

Across the show floor, exhibitors and attendees talked about ways the upstream oil and gas industry can think differently and optimize operations to succeed in a low-price market.

"You have to provide product that fits the changing industry. Reserve Energy Exploration has done that, and we've had a good show," said Joe Haas,

President of Reserve Energy Exploration.

"We're focused on providing integrated solutions that help E and Ps streamline operations and do business more efficiently, a necessity in today's new price environment," said Nicole Durham, Marketing Director for Enertia Software.

This year, Summer NAPE featured several new attractions to help attendees build their careers and their businesses, including the new Capital and Service Provider Presentations, which showcased new technologies, tools and data analytics software available for landmen and operators.

Among the most popular of Summer NAPE's new attractions was its free professional headshots booth, which had a line wrapping around the booth for most of the day.

George Oggero, RPL with P.O.andG. Resources, LP had this to say regarding the free professional headshots, "I didn't know Summer NAPE would be offering free headshots. It was a nice surprise this year".

Summer NAPE 2016 wrapped up with the Closing Celebration Reception sponsored by Bounty Minerals.

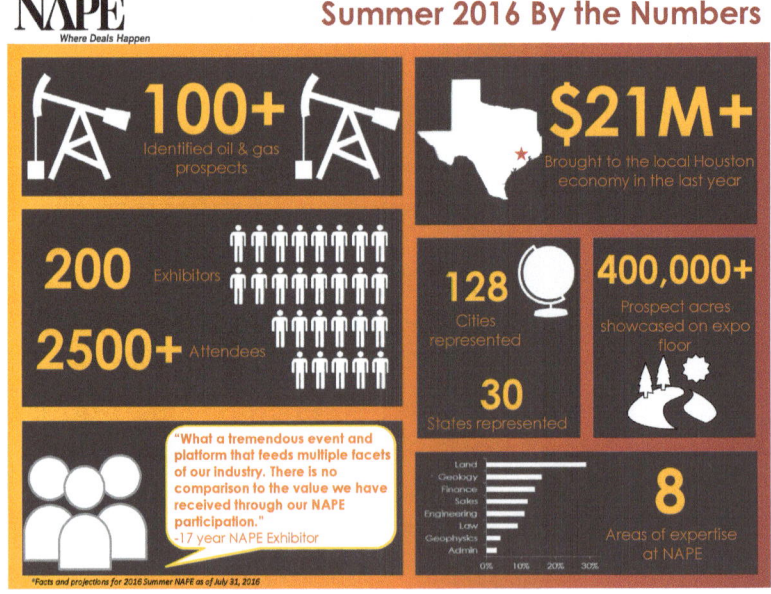

Damen USA office off to a good start with Young Brothers' order for Damen of 4 Stan Tugs 3711

Damen Shipbuilding Group is establishing a permanent presence in Houston, Texas. The office opened on 1st August and will be headed by senior managers Jan van Hogerwou New Construction and Ruud Haneveer Ship Repair & Conversion. This move is part of Damen's policy of expanding its local footprint around the world, bringing it closer to current and prospective clients and enabling it to deliver a more responsive and personal service.

The first order to be handled by the new entity is for four new Damen 3711 Stan Tugs by Young Brothers, Limited, Hawaii's largest inter-island cargo service provider. Young Brothers is a Saltchuk company operated by Foss Maritime. The tugs will be built at Conrad Shipyard, Louisiana, under a license and materials agreement with Damen. The first vessel will be delivered in the first quarter of 2018 and the last twelve months later. Together the 37-meter, 6,000 hp tugs will service Young Brothers' fleet of modern, high-capacity tugs & barges.

"Opening a permanent Damen presence in North America is a significant milestone for the group," said Jan van Hogerwou. "With over 200 Damen design vessels built and delivered in the US, this has been an important market for our vessels for many years via our flexible licensing agreements, and we have enjoyed excellent cooperation with shipyards across the country. This latest initiative will enable us to strengthen our relationships further with both builders and operators, and serve the North American market with the innovative, cost-effective and dependable vessels that it seeks."

The choice of Houston as the location of the new office is also intended to send a message to the offshore oil & gas sector in North America. That is, that Damen is committed to supporting and working with vessel owners and operators to maximise efficiencies and minimise costs, despite the current downturn. It will support Damen's work with shipyards and owners right across the maritime sector and Damen's product range. The new Damen office is registered in Houston Texas as Damen Area Support North America BV Co, and can be contacted via 954 591 6139.